Cat Musculature
A Photographic Atlas

Second Edition

Gordon M. Greenblatt

The University of Chicago Press
Chicago and London

1

The University of Chicago Press, Chicago 60637
The University of Chicago Press, Ltd., London
©1954, 1980 by The University of Chicago
All rights reserved. Second edition published 1980
Printed in the United States of America

88 87 86 85 84 83 82 81 80   5 4 3 2 1

Library of Congress Cataloging in Publication Data

Greenblatt, Gordon M   1932–
    Cat musculature.

    Updates the page references to the 3d ed. of Hyman's
Comparative vertebrate anatomy.
    1. Muscles.   2. Cats—Anatomy.   I. Hyman, Libbie
Henrietta, 1888–1969. Comparative vertebrate anatomy.
II. Title.
QL812.G76 1980 599.74'428   80-25610
ISBN 0-226-30656-9

Preface

This series of photographs is meant to supplement the descriptions of cat musculature found in *Hyman's Comparative Vertebrate Anatomy*, third edition, edited by Marvalee Wake, or any other good comparative anatomy text. It is not to be used as the sole text for the laboratory, but only to assist the student in locating muscles with the aid of instructions and fuller descriptions in a basic text.

At the end of the list of muscles on each page of this atlas are the page numbers in the Wake text on which those muscles are discussed. The page references to the Wake text are also listed in the index for this atlas, together with information on origin, insertion, and action.

These photographs should make the descriptions in Wake or any other comparative anatomy text more vivid and clear, for the student can use these photographs to see what the muscle looks like before starting dissection.

All dissection and photography were done by Gordon M. Greenblatt.

Guide to

CAT MUSCULATURE: A PHOTOGRAPHIC ATLAS

This guide coincides with the organization of the material in *Hyman's Comparative Vertebrate Anatomy*, third edition, edited by Marvalee Wake. Information on the origins, insertions, and actions of the muscles are also from Wake. Numbers following the names of individual muscles refer to plates in this atlas.

Pectoral Girdle and Forelimb Muscles

Superficial Chest Muscles (pages 355 and 356 in Wake)

1. Pectoantibrachialis 2, 15
   Origin: Manubrium
   Insertion: Fascia of forearm
   Action: Draws arm towards chest

2. Pectoralis major 2, 15
   Origin: Sternum and median ventral raphe
   Insertion: Pectoral ridge on ventral side of humerus
   Action: Draws arm towards chest

3. Pectoralis minor 2, 15
   Origin: Sternum
   Insertion: Ventral side of humerus, distal to pectoralis major
   Action: Drawn arm toward chest

4. Xiphihumeralis 2, 15
   Origin: Xiphoid process
   Insertion: Ventral side of humerus with pectoralis minor
   Action: Draws arm towards chest

Shoulder Muscles (pages 356 to 359 in Wake)

5. Latissimus dorsi 3
   Origin: Neural spines of last thoracic and lumbar vertebrae and lumbodorsal fascia
   Insertion: Medial surface of humerus
   Action: Pulls forelimb dorsally and caudally, by retracting humerus

Trapezius muscles

6. Spinotrapezius 3
   Origin: Spines of thoracic vertebrae
   Insertion: Fascia of scapula
   Action: Draws scapula dorsad and caudad

7. Acromiotrapezius 3
   Origin: Spines of cervical and first thoracic verteb[rae]
   Insertion: Metacromion process, spine of scapula, fascia of spinotrapezius
   Action: Draws scapula dorsad and holds scapulae together

8. Clavotrapezius 3
   Origin: Superior nuchal line and median dorsal line [of] neck
   Insertion: Clavicle
   Action: Draws clavicle dorsad and craniad

9. Levator scapulae ventralis 3
   Origin: Transverse process of atlas and occipital bon[e]
   Insertion: Metacromion process of scapula and neighbor[ing] fascia
   Action: Draws scapula craniad

10. Rhomboideus 4
    Origin: Neural spines of vertebrae and ligaments
    Insertion: Vertebral border of scapula
    Action: Draws scapula dorsad and medial

11. Rhomboideus capitis 4
    Origin: Superior nuchal line
    Insertion: Scapula
    Action: Draws scapula craniad or raises head

Deltoids

12. Clavodeltoid (clavobrachialis) 3
    Origin: Clavicle and fibers of clavotrapezius
    Insertion: Ulna
    Action: Flexor of forearm or protracts arm

13. Acromiodeltoid 3
    Origin: Acromion process
    Insertion: Spinodeltoid and brachial muscles and humeral deltoid ridge
    Action: Retracts and abducts humerus

14. Spinodeltoid 3
    Origin: Scapular spine
    Insertion: Deltoid ridge of humerus
    Action: Retracts and abducts humerus

15. Supraspinatus 4

Origin: Surface of supraspinous fossa
Insertion: Greater tuberosity of humerus
Action: Extends humerus

Infraspinatus 4
Origin: Infraspinous fossa of scapula
Insertion: Greater tuberosity of humerus
Action: Rotates humerus outward

Teres major 5
Origin: Glenoid border of scapula and neighboring fascia
Insertion: Medial surface of humerus with latissimus dorsi
Action: Rotates and retracts humerus

Teres minor 5
Origin: Glenoid border of scapula
Insertion: Greater tuberoisty of humerus
Action: Rotates humerus with infraspinatus

Subscapularis 6
Origin: Subscapular fossa
Insertion: Lesser tuberosity of humerus
Action: Pulls humerus medially

Serratus ventralis 6
Origin: First nine or ten ribs and transverse processes of last five cervical vertebrae
Insertion: Scapula near vertebral border
Action: Anterior part draws scapula craniad, main part draws scapula ventrad

pper Arm Muscles (pages 359 to 360 in Wake)

Triceps brachii long head 5
Origin: Scapula, glenoid border
Insertion: Olecranon process
Action: With lateral and medial heads, extensor of forearm

Triceps brachii lateral head 5
Origin: Greater tuberosity and deltoid ridge of humerus
Insertion: Olecranon process
Action: With long and medial heads, extensor of forearm

Triceps brachii medial head 5
Origin: Dorsal surface of humerus
Insertion: Olecranon process
Action: With long and lateral heads, extensor of forearm

Triceps brachii fourth head (anconeus) 5
Origin: Distal end of humerus

Insertion: Lateral surface of ulna
Action: Strengthens elbow joint

Epitrochlearis 7
Origin: Latissimus dorsi
Insertion: Olecranon process
Action: Extensor of forearm and rotator of ulna

Biceps brachii 7
Origin: Supraglenoid tubercle of scapula
Insertion: Bicipital tuberosity of radius
Action: Flexes and supinates forearm

Brachialis 5
Origin: Lateral surface of humerus
Insertion: Ulna
Action: With the biceps

Forearm Muscles (pages 360 and 361 in Wake)

Extensor carpi ulnaris 8
Origin: Lateral epicondyle of humerus and semilunar notch of ulna
Insertion: Proximal end of fifth metacarpal
Action: Extends fifth digit and ulnar side of wrist

Extensor digitorum lateralis 8
Origin: Lateral surface of humerus above lateral epicondyle
Insertion: Tendons to three or four digits
Action: Extends digits

Extensor digitorum communis 8
Origin: Lateral surface of humerus above lateral epicondyle
Insertion: Tendons to three or four digits
Action: Extends digits

Brachioradialis (supinator longus) 8
Origin: Middle of humerus
Insertion: Lower end of radius and ligaments
Action: Rotates hand to supine position

Extensor carpi radialis 8, 9
Origin: Humerus
Insertion: Second and third metacarpals
Action: Extends hand

Pronator teres 9
Origin: Medial epicondyle of humerus

Insertion: Radius
Action: Rotates radius to prone position
Flexor carpi radialis 9
   Origin: Medial epicondyle of humerus
   Insertion: Second and third metacarpals
   Action: Flexes hand
Palmaris longus 9
   Origin: Medial epicondyle of humerus
   Insertion: Tendons to pads of palm and digits
   Action: Flexor of digits
Flexor digitorum profundus 10
   Origin: Ulna and humerus
   Insertion: Basal phalanges
   Action: Flexor of digits
Flexor carpi ulnaris 9
   Origin: Medial epicondyle of humerus and olecranon process
   Insertion: Pisiform bone
   Action: Flexes ulnar side of wrist

Trunk Muscles

Hypaxial Muscles (pages 361 and 362 in Wake)
  External oblique 11
   Origin: Lumbodorsal fascia and posterior ribs
   Insertion: Aponeurosis to linea alba
   Action: Constrictor of abdomen
  Internal oblique 11
   Origin: Second sheet of lumbodorsal fascia and border of pelvic girdle
   Insertion: Aponeurosis to linea alba
   Action: Compressor of abdomen
  Transverse 11
   Origin: Second sheet of lumbodorsal fascia and border of pelvic girdle
   Insertion: Aponeurosis to linea alba
   Action: Compressor of abdomen
  Rectus abdominis 11
   Origin: Anterior end of pubic symphysis

Insertion: Sternum and costal cartilages
Action: Retracts ribs and sternum and compresses abdomen
Serratus dorsalis 4
   Origin: Aponeurosis to medial dorsal line
   Insertion: Ribs near angles
   Action: Draws ribs forward
Scalenes 6
   Origin: Ribs
   Insertion: Transverse processes of cervical vertebrae
   Action: Draw ribs forward and bend neck
Epaxial muscles (pages 362 and 363 in Wake)
  Sacrospinalis 12
   Semispinalis dorsi
    Action: Dorsiflex the back
   Longissimus
    Action: Dorsiflex the back
   Iliocostalis
    Action: Draws ribs together
27 Splenius 12
   Origin: Middorsal line and fascia
   Insertion: Superior nuchal line
   Action: Raises or turns head
  Longissimus capitis 12
   Origin: Last four cervical vertebrae
   Insertion: Mastoid process
   Action: Turns head

Head and Neck Muscles (pages 363 to 366 in Wake)

28 Sternomastoid 13, 15
   Origin: Median raphe and manubrium of sternum
   Insertion: Mastoid process and lamboidal ridge
   Action: Singly turns head, together depress head on neck
29 Sternohyoid 13, 15
   Origin: First costal cartilage
   Insertion: Hyoid bone
   Action: Draws hyoid posteriorly
30 Cleidomastoid 13, 15

Origin: Clavicle
Insertion: Mastoid process (origin and insertion are
  interchangeable)
Action: Singly pulls clavicle craniad or turns head,
  together lower head on neck
Masseter   13, 15   *31*
  Origin: Zygomatic arch
  Insertion: Posterior half of lateral mandibular surface
  Action: Elevator of jaw
Temporalis   14   *32*
  Origin: Superior nuchal line to zygomatic arch
  Insertion: Coronoid process of mandible
  Action: Elevator of jaw with masseter
Digastric   13, 15   *33*
  Origin: Jugular and mastoid process of skull
  Insertion: Mandible
  Action: Depresses lower jaw
Mylohyoid   13, 15   *34*
  Origin: Mandible
  Insertion: Median raphe
  Action: Raises floor of mouth and brings hyoid bone
    forward
Geniohyoid   13, 15
  Origin: Mandible near symphysis
  Insertion: Body of hyoid bone
  Action: Draws hyoid forward
Sternothyroid   13, 15
  Origin: Sternum
  Insertion: Thyroid cartilage
  Action: Pulls larynx posteriorly
Thyrohyoid   13, 15
  Origin: Posterior horn of hyoid
  Insertion: The same
  Action: Raises the larynx

Pelvic Girdle and Hind Limb Muscles

Hip and Thigh Muscles (pages 366 to 370 in Wake)
  Tensor fascie latae   16
    Origin: Ilium and fascia
    Insertion: Fascia lata
    Action: Tightens fascia lata
  Biceps fermoris   16
    Origin: Tuberosity of ischium
    Insertion: Patella, tibia, and shank fascia
    Action: Abductor of thigh and flexor of shank
  Caudofemoralis   16
    Origin: Transverse processes of second and third caudal
      vertebrae
    Insertion: Patella
    Action: Abductor of thigh, extensor of shank
  Gluteus maximus   16
    Origin: Fascia and transverse processes of last sacral and
      first caudal vertebrae
    Insertion: Fascia lata and greater trochanter
    Action: Abductor of thigh
  Gluteus medius   16
    Origin: Fascia, crest, and lateral surface of ilium, and
      transverse processes of last sacral and first caudal
      vertebrae
    Insertion: Greater trochanter
    Action: Abductor of thigh
  Sartorius   17
    Origin: Crest and ventral border of ilium
    Insertion: Proximal end of tibia, patella, and fascia
    Action: Adductor and rotator of thigh and extensor of
      shank
  Vastus lateralis   19
    Origin: Greater trochanter and surface of femur
    Insertion: Patella
    Action: Extensor of shank
  Rectus femoris   18
    Origin: Ilium near acetabulum
    Insertion: Patella
    Action: Extensor of shank

Vastus medialis   18
  Origin: Femur
  Insertion: Patella
  Action: Extensor of shank
Vastus intermedius   20
  Origin: Surface of femur
  Insertion: Patella
  Action: Extensor of shank
Gracilis   17
  Origin: Ischial and pubic symphysis
  Insertion: Aponeurosis to tibia
  Action: Adductor of leg
Adductor longus   18
  Origin: Pubis
  Insertion: Femur
  Action: Adductor of leg
Adductor femoris   18
  Origin: Pubis
  Insertion: Femur
  Action: Adductor of leg
Semimembranosus   18
  Origin: Ischium
  Insertion: Medial epicondyle of femur and proximal end of tibia
  Action: Extensor of thigh
Semitendinosus   19
  Origin: Ischial tuberosity
  Insertion: Tibia
  Action: Flexor of shank
Tenuissimus   19
  Origin: Transverse processes of second caudal vertebrae
  Insertion: On same fascia as insertion of biceps
  Action: With biceps
Muscles of the Shank (pages 370 and 371 in Wake)
  Tibialis anterior   21
    Origin: Proximal parts of tibia and fibula
    Insertion: First metatarsal
    Action: Flexor of foot
  Extensor digitorum longus   22
    Origin: Lateral epicondyle of femur

Insertion: Digits
Action: Extensor of digits and dorsiflexion of foot
Peroneus   22
  Origin: Fibula
  Insertion: Metatarsals and digits
  Action: Extensor of foot
Gastrocnemius   21, 22
  Origin: Surface fascia, femur, tendon, and fascia of plantaris
  Insertion: Calcaneus
  Action: Extensor of foot
Soleus   22
  Origin: Fibula
  Insertion: Calcaneus
  Action: Extensor of foot
Plantaris   23
  Origin: Patella and femur
  Insertion: Ventral surface of calcaneus and digits
  Action: Flexor of digits and extensor of ankle
Flexor digitorum longus   21
  Origin: Tibia, fibula, and fascia
  Insertion: Digits
  Action: Flexor of digits
Tibialis posterior   24
  Origin: Fibula, tibia, and fascia
  Insertion: Scaphoid and medial cuneiform of ankle
  Action: Extensor of foot

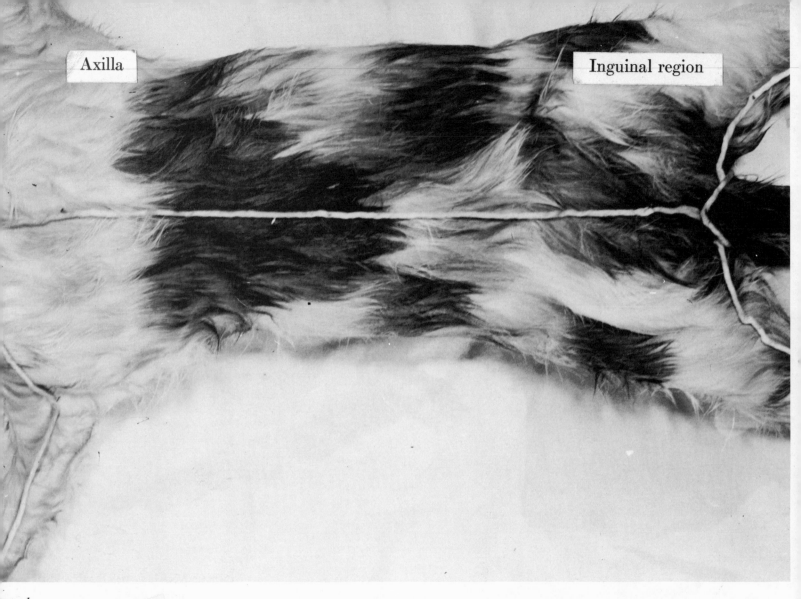

Axilla

Inguinal region

*ate 1
orsal view, showing locations of
inning incisions
age 354 in Wake)

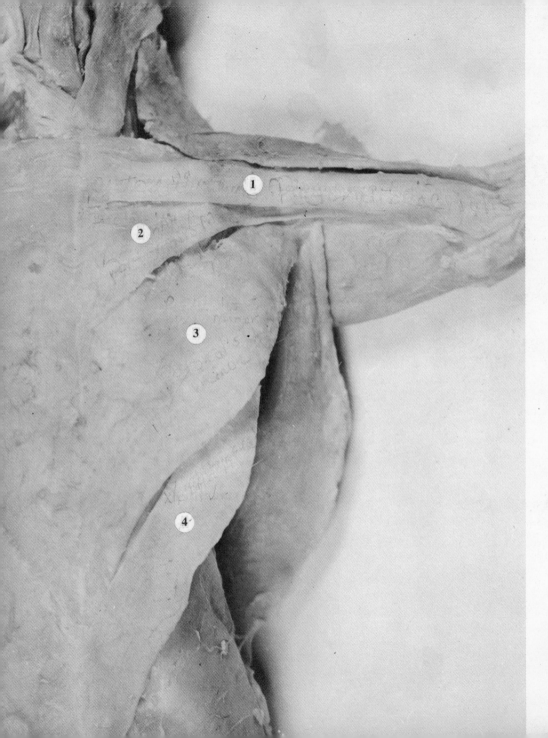

Plate 2
SUPERFICIAL CHEST MUSCLES
1. Pectoantibrachialis
2. Pectoralis major
3. Pectoralis minor
4. Xiphihumeralis
(Page 355 in Wake)

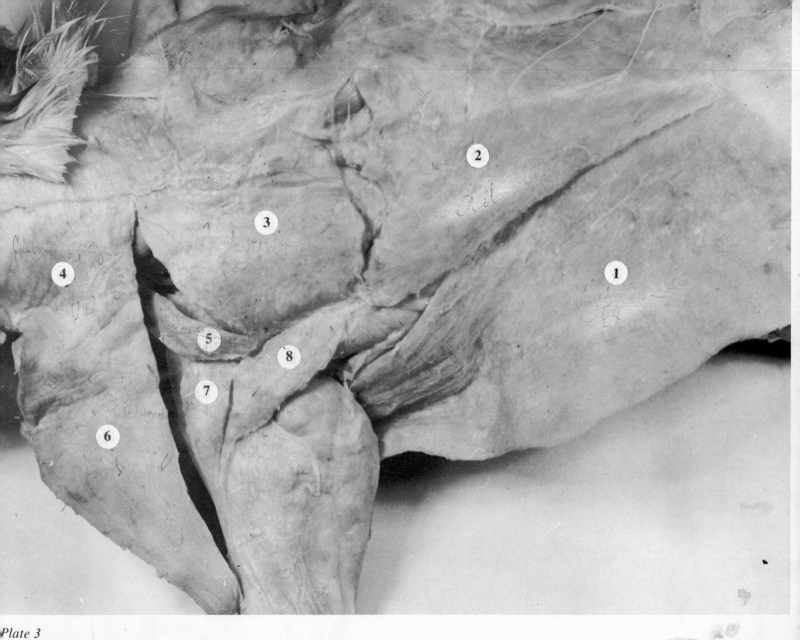

Plate 3
SHOULDER MUSCLES
1. Latissimus dorsi
2. Spinotrapezius
3. Acromiotrapezius
4. Clavotrapezius
5. Levator scapulae ventralis
6. Clavodeltoid
7. Acromiodeltoid
8. Spinodeltoid
(Page 356 in Wake)

11

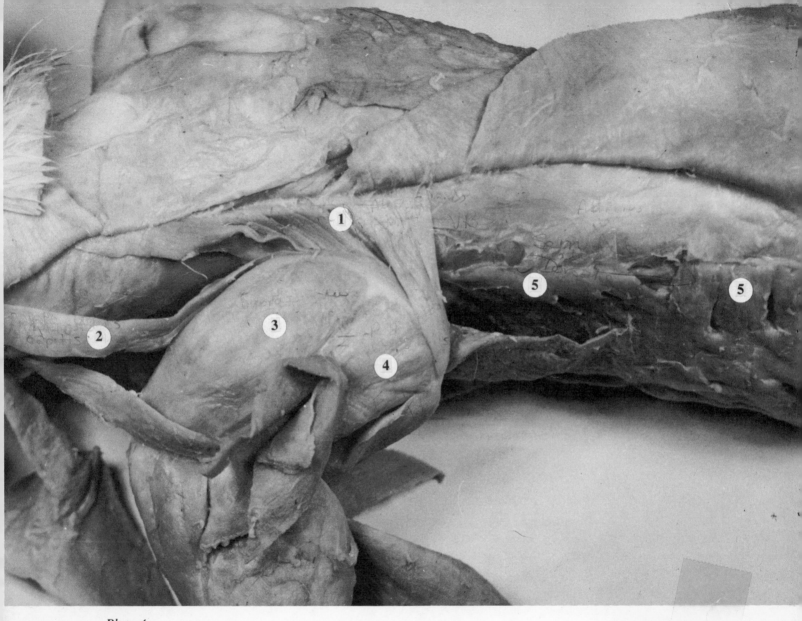

Plate 4
SHOULDER MUSCLES   4. Infraspinatus
1. Rhomboideus   TRUNK MUSCLES
2. Rhomboideus capitis   5. Serratus dorsalis
3. Supraspinatus   (Pages 357, 358, and 362 in Wake)

Plate 5
SHOULDER MUSCLES   UPPER ARM MUSCLES
1. Teres major   3. Triceps brachii, long head
2. Teres minor   4. Triceps brachii, lateral head
    5. Triceps brachii, medial head

6. Triceps brachii, fourth head (anconeus)
7. Brachialis
(Pages 358 and 359 in Wake)

13

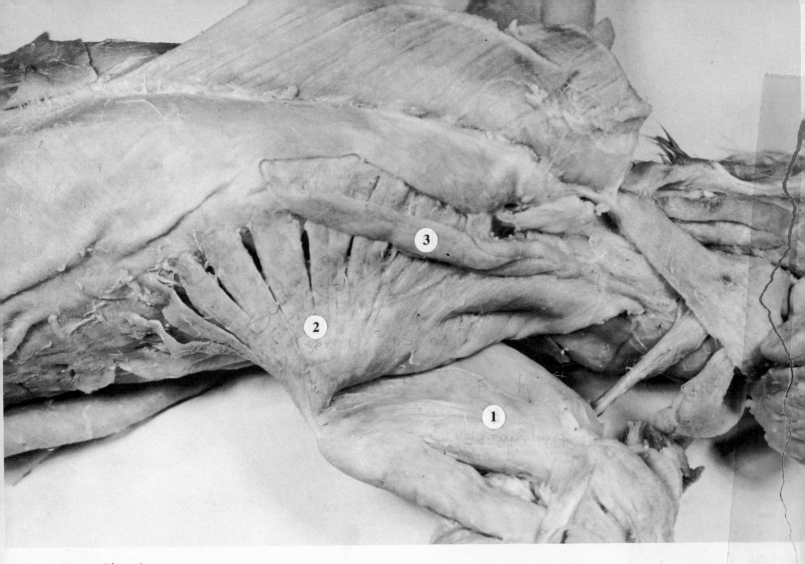

Plate 6
SHOULDER MUSCLES
1. Subscapularis
2. Serratus ventralis
TRUNK MUSCLES
3. Scalenes
(Pages 358, 359, and 362 in Wake)

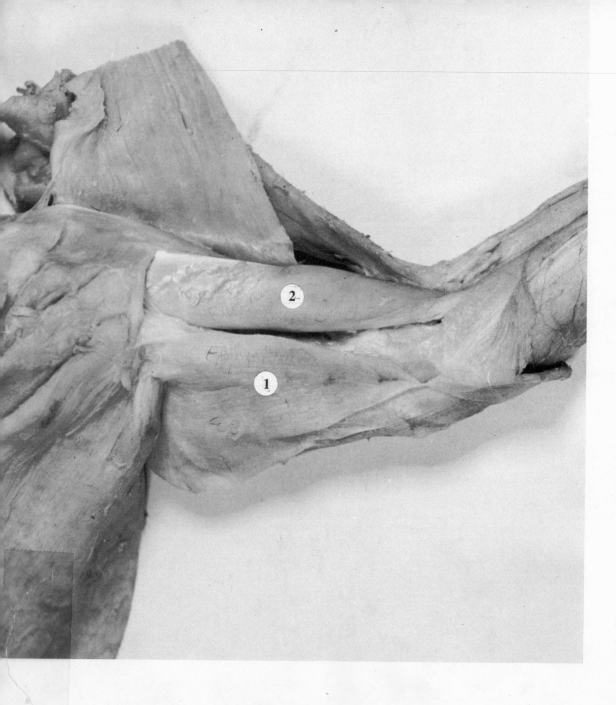

Plate 7
UPPER ARM MUSCLES
1. Epitrochlearis
2. Biceps brachii
(Page 359 in Wake)

15

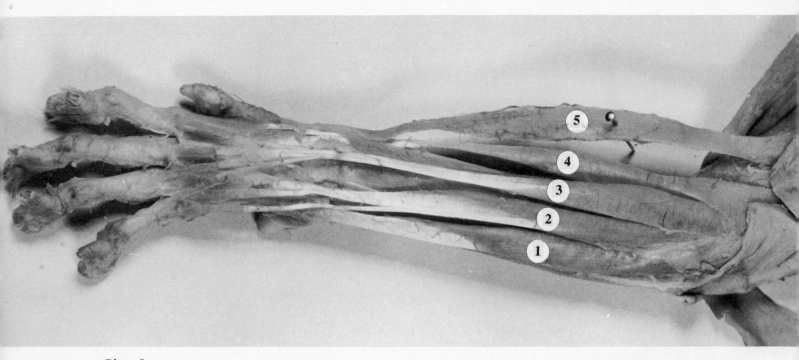

Plate 8
FOREARM MUSCELS
1. Extensor carpi ulnaris
2. Extensor digitorum lateralis
3. Extensor digitorum communis
4. Brachioradialis (supinator longus)
5. Extensor carpi radialis
(Page 360 in Wake)

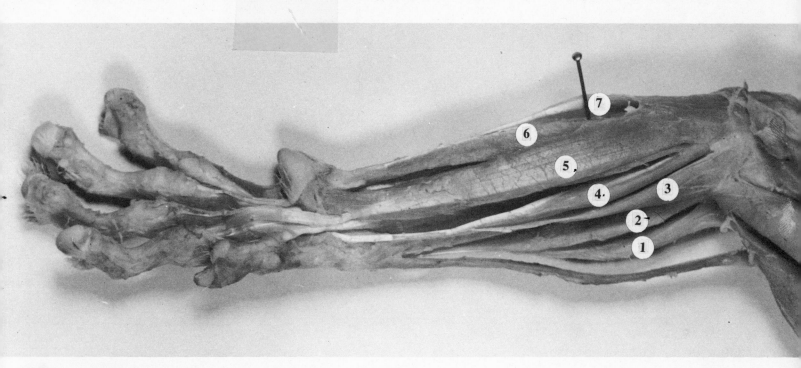

Plate 9

FOREARM MUSCLES

1. Extensor carpi radialis longus
2. Extensor carpi radialis brevis
3. Pronator teres
4. Flexor carpi radialis
5. Palmaris longus
6. Flexor carpi ulnaris, first head
7. Flexor carpi ulnaris, second head
(Page 360 in Wake)

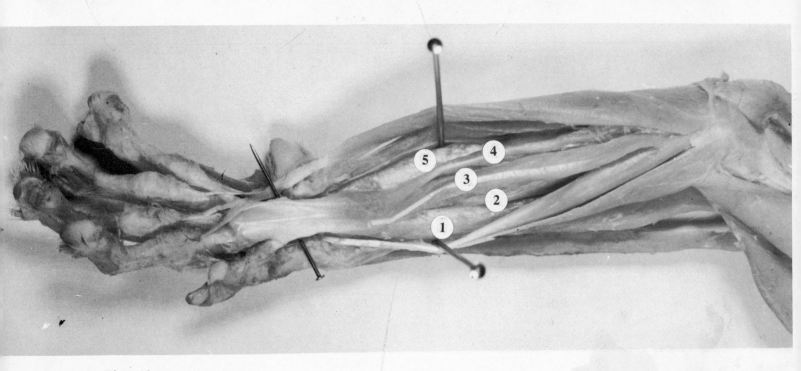

Plate 10

FOREARM MUSCLES

1. Flexor digitorum profundus, first head
2. Flexor digitorum profundus, second head
3. Flexor digitorum profundus, third head
4. Flexor digitorum profundus, fourth head
5. Flexor digitorum profundus, fifth head

(Page 361 in Wake)

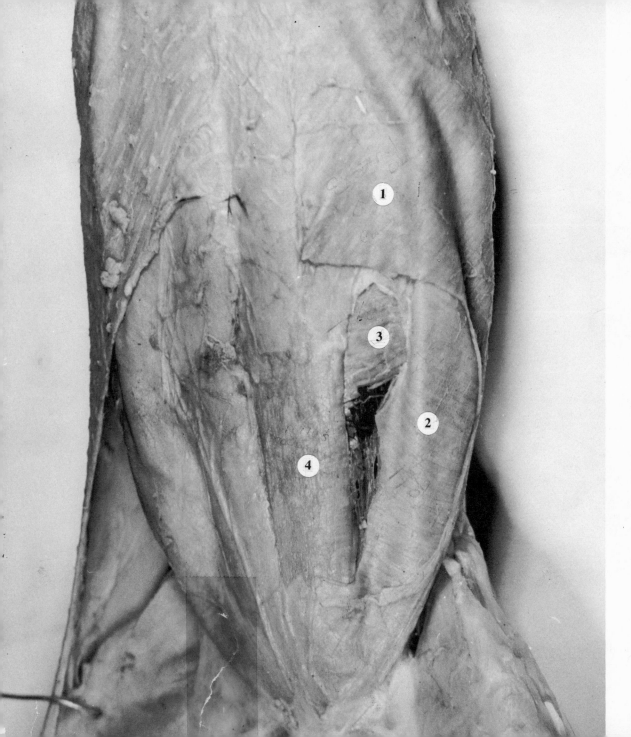

Plate 11
HYPAXIAL MUSCLES
1. External oblique
2. Internal oblique
3. Transverse
4. Rectus abdominis
(Page 361 in Wake)

19

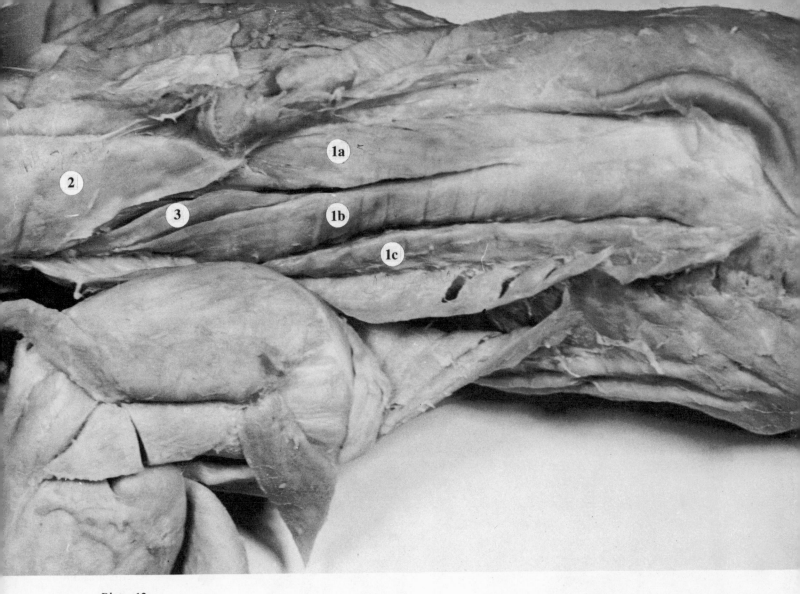

Plate 12

EPAXIAL MUSCLES

1. Sacrospinalis
    a. Semispinalis dorsi
    b. Longissimus
    c. Iliocostalis

2. Splenius
3. Longissimus capitis
(Pages 362 and 363 in Wake)

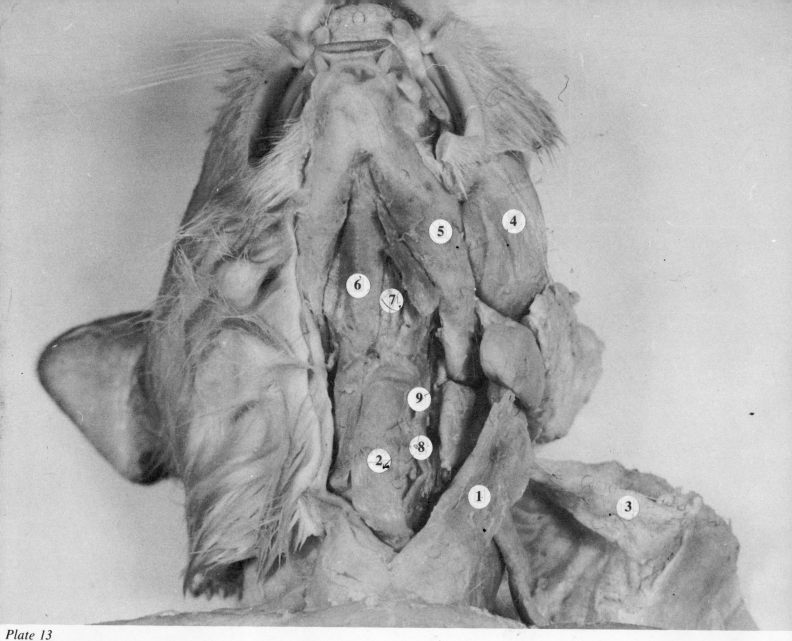

Plate 13

HEAD AND
NECK MUSCLES

1. Sternomastoid
2. Sternohyoid

3. Cleidomastoid
4. Masseter
5. Digastric

6. Mylohyoid
7. Geniohyoid
8. Sternothyroid

9. Thyrohyoid
(Pages 363 and 364 in Wake)

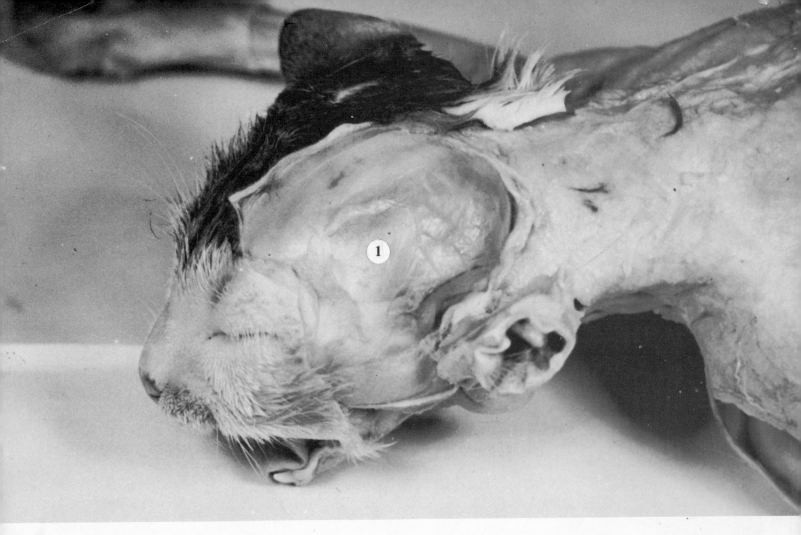

Plate 14
HEAD MUSCLES
1. Temporalis
(Page 364 in Wake)

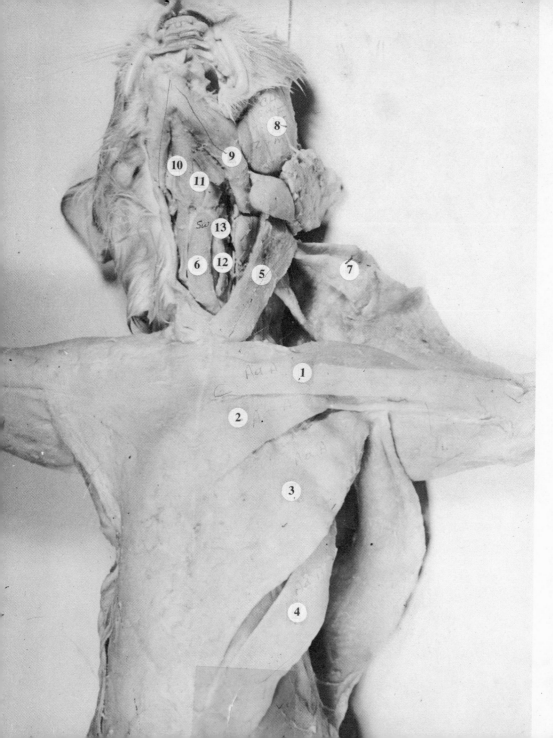

Plate 15

SUPERFICIAL CHEST MUSCLES
1. Pectoantibrachialis
2. Pectoralis major
3. Pectoralis minor
4. Xiphihumeralis

HEAD AND NECK MUSCLES
5. Sternomastoid
6. Sternohyoid
7. Cleidomastoid
8. Masseter
9. Digastric
10. Mylohyoid
11. Geniohyoid
12. Sternothyroid
13. Thyrohyoid

(Pages 355 and 363 in Wake)

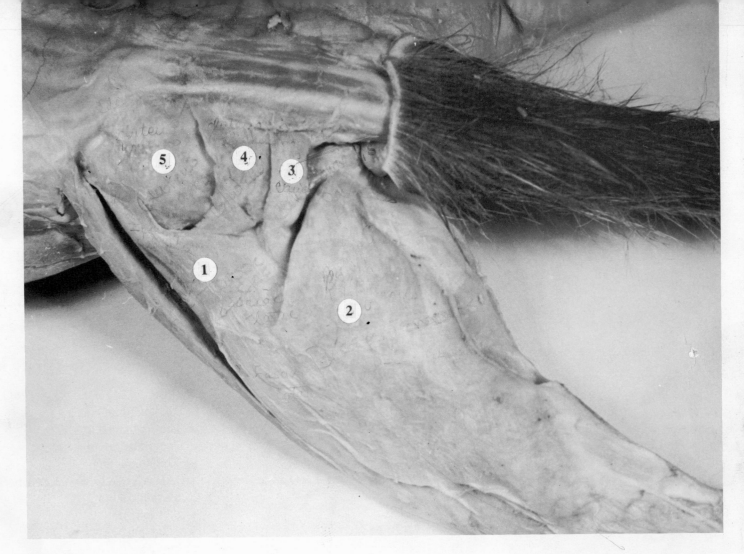

Plate 16

HIP AND THIGH MUSCLES

1. Tensor fasciae latae
2. Biceps femoris
3. Caudofemoralis
4. Gluteus maximus
5. Gluteus medius
(Pages 366, 367, and 368 in Wake)

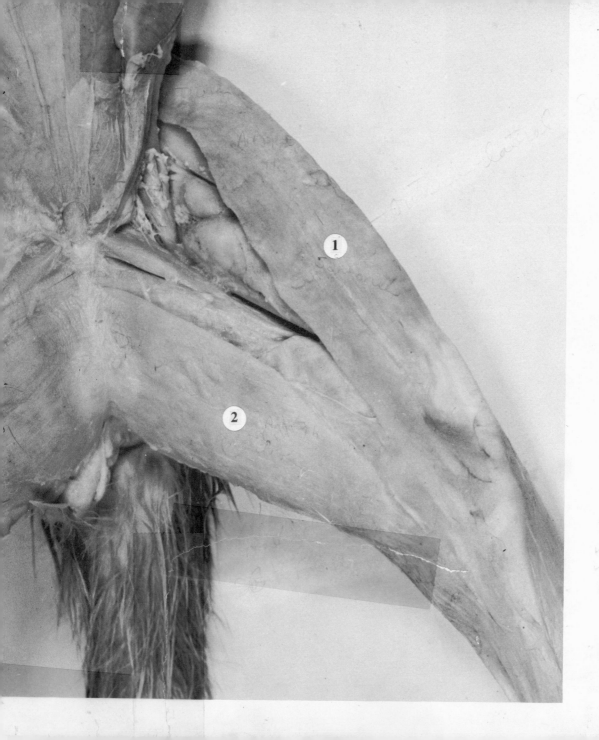

Plate 17
THIGH MUSCLES
1. Sartorius
2. Gracilis
(Pages 368 and 369 in Wake)

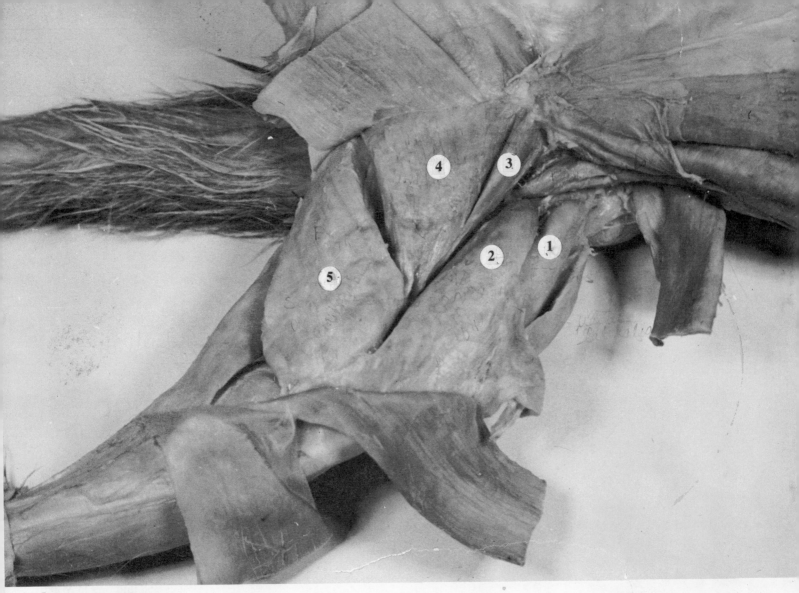

Plate 18
THIGH MUSCLES
1. Rectus femoris
2. Vastus medialis
3. Adductor longus
4. Adductor femoris
5. Semimembranosus
(Pages 368 and 369 in Wake)

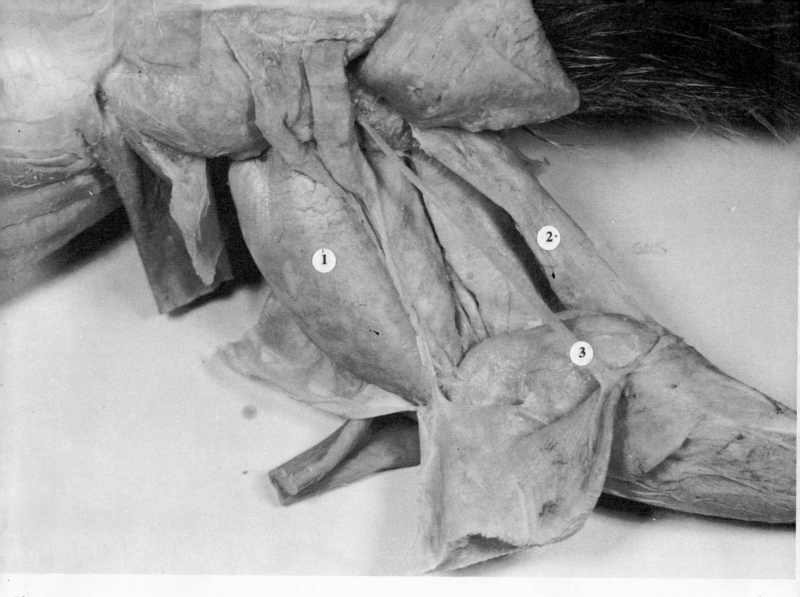

Plate 19
THIGH MUSCLES
1. Vastus lateralis
2. Semitendinosus
3. Tenuissimus
(Pages 368 and 369 in Wake)

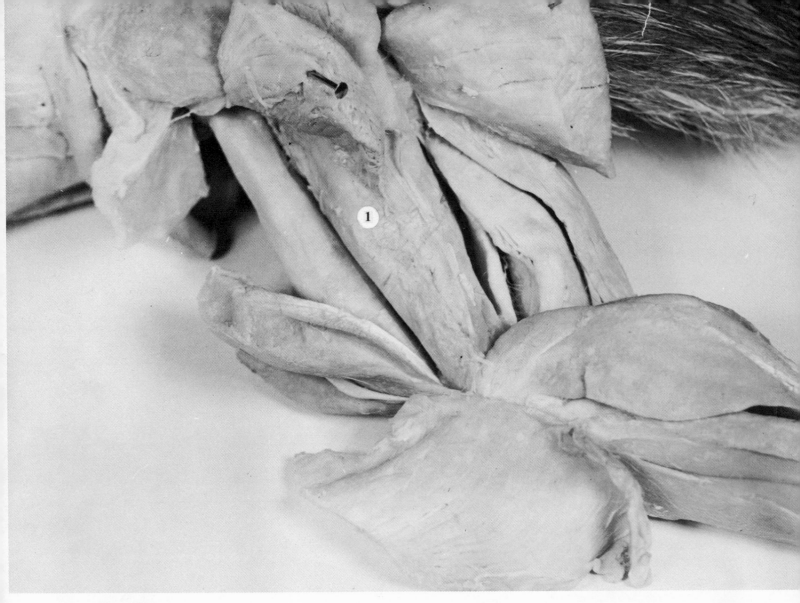

Plate 20
THIGH MUSCLES
1. Vastus intermedius
(Page 368 in Wake)

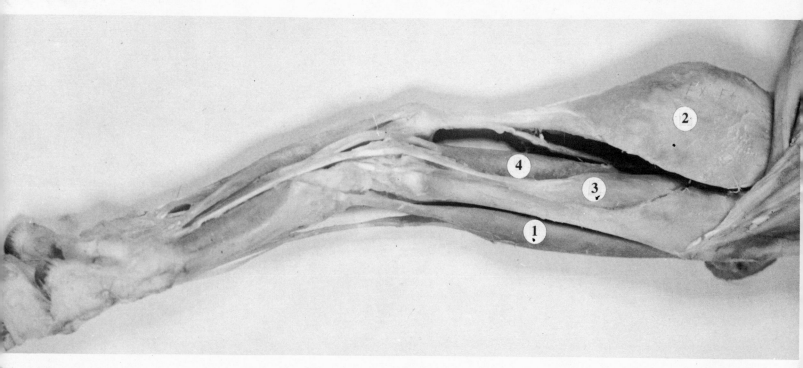

Plate 21
SHANK MUSCLES
1. Tibialis anterior
2. Gastrocnemius, medial head
3. Flexor digitorum longus, first head
4. Flexor digitorum longus, second head
(Pages 370 and 371 in Wake)

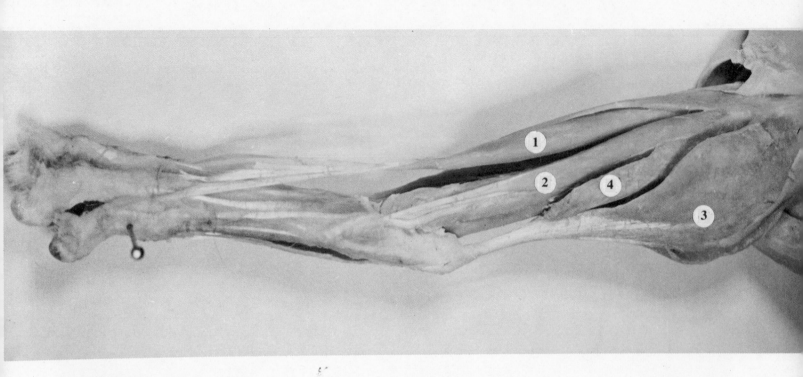

Plate 22
SHANK MUSCLES
1. Extensor digitorum longus
2. Peroneus
3. Gastrocnemius, lateral head
4. Soleus
(Page 370 in Wake)

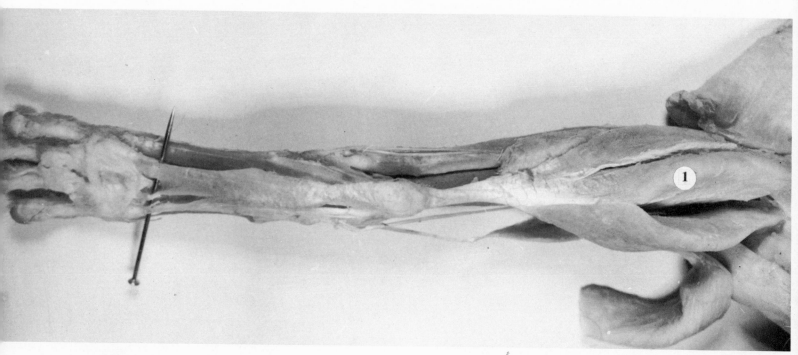

Plate 23
SHANK MUSCLES
1. Plantaris
(Page 371 in Wake)

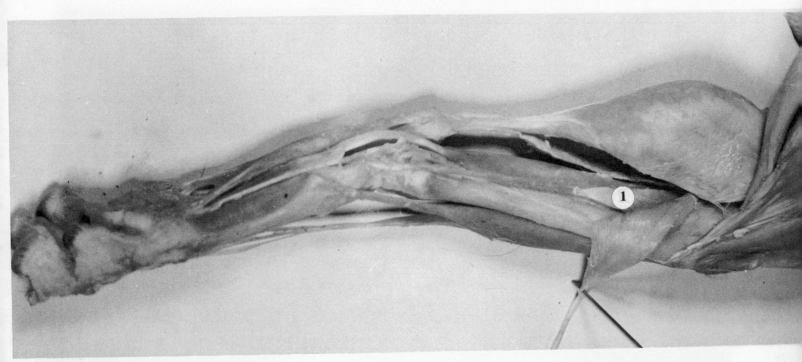

Plate 24
SHANK MUSCLES
1. Tibialis posterior
(Page 371 in Wake)